# THE FULL-SPECTRUM HEALER

The ornamental ginkgo tree, decorating city streets and parks, is now known to possess a wide range of therapeutic properties and to act as a powerful scavenger of body-damaging free radicals. Frank Murray details studies showing ginkgo extract's curative effects in cases of Alzheimer's disease, headaches, circulatory disorders, hemorrhoids, eye problems, ringing in the ears and asthma—and why it is one of our most potent antioxidants.

# ABOUT THE AUTHOR

Frank Murray, editorial director of *Let's Live* magazine, is the author or coauthor of 30 books on health and nutrition, the most recent being *The Big Family Guide to All the Minerals*.

# GINKGO BILOBA

*Therapeutic and antioxidant properties of the "tree of health"*

# Frank Murray

Keats Publishing, Inc.  New Canaan, Connecticut

GINKGO BILOBA

Copyright © 1996 by Frank Murray

All Rights Reserved

No part of this book may be reproduced in any form without the written consent of the publisher.

Library of Congress Cataloging-in-Publication Data

Murray, Frank, 1924–
    Ginkgo Biloba / Frank Murray.
        p.     cm.—(A Keats good herb guide)
    Includes bibliographical references and index.
    ISBN 0-87983-770-5
    1. Ginkgo—Therapeutic use.   I Title.   II. Series.
    RM666.G489M87   1996
    615'.3257—dc20                                              96-24280
                                                                        CIP

Printed in the United States of America

Keats Good Health Guides are published by
Keats Publishing, Inc.
27 Pine Street (Box 876)
New Canaan, Connecticut 06840-0876

98  97  96      6  5  4  3  2  1

# CONTENTS

# INTRODUCTION

The forests, swamps and grasslands of the world abound with plants endowed with the power to heal almost all the ailments humans experience. Researchers believe that there remain undiscovered many medicinal herbs that might be even more striking in their curative effects than those we know of now— the inventory of the "wild pharmacy" is not nearly complete.

But one of the most effective of all medicinal plants is nowhere to be found in the wild. The sole survivor of its ancient family, it grows only where humans, not nature, have planted it.

Fortunately, the ginkgo tree's appeal as an ornamental plant has led to its widespread use throughout the world—fortunately, because ginkgo has powerful therapeutic properties for a number of serious medical conditions, including Alzheimer's disease, asthma, impotence, multiple sclerosis, tinnitus and hearing loss, headaches, circulatory disorders and hemorrhoids.

## The "Living Fossil"

A native of southeastern China, and a popular ornamental tree in many parts of the world, ginkgo biloba is a member of the Ginkgoales family, which dates from the Permian period of the Paleozoic era between

225 and 280 million years ago. It is the last surviving member of that family and is regarded as a living fossil, since it cannot be found in the wild.

Ginkgo (pronounced GIHNG-koh), also known as the maidenhair tree, is a stately deciduous tree 100 feet or more in height, with a trunk that can extend to eight feet in diameter. It was widely distributed over the temperate regions of both the Northern and Southern Hemisphere during the time of the dinosaurs.

The fan-shaped, leathery leaves of the ginkgo, from two to three inches across and the source of the tree's medicinal wonders, are not found on any other flowering plant, although they somewhat resemble the leaflets of the maidenhair fern. Since the two share several common features, ginkgo is sometimes referred to as the "maidenhair tree," and botanists believe that it is a missing link between flowering plants and ferns.

The word ginkgo is derived from the Japanese *ginkyo* and the Chinese *yinhsing,* which translates as "silver apricot." The leaves of the tree are divided into two lobes, hence *biloba.*

The fissured bark of the ginkgo is grayish and deeply furrowed on older trees and has a corky texture. The leaves, generally divided into two lobes by a central notch, range from dull gray-green to yellow-green during the summer. They turn golden yellow in the autumn and eventually fall off.

There are both male and female ginkgo trees. Male flowers look something like catkins (a cat's tail) and are borne on separate trees from the female flowers. The male flowers produce free-swimming reproductive cells, which is characteristic of ferns and cycads (tropical plants such as palms), but not of gnetums or conifers (cone-bearing plants). The male cells are delivered to female trees by the wind.

Female trees produce paired ovules which, when fertilized, develop into yellowish, plumlike seeds about one inch long. The seed consists of a large silvery nut surrounded by a fleshy outer covering and, when ripe, smells something like rancid butter.

The kernels from the female trees, called ginkgo, are sold in China, Japan and other countries in Southeast Asia. When roasted, they are considered a delicacy.

Since male ginkgo trees do not produce the foul-smelling seeds, they are generally preferred as ornamentals, so it is important to know the sex of the ginkgo tree you are planting. The large, hard kernel of the fruit from the female tree can cause pedestrians to slip and fall. If clothing is stained by the juice, it generally has to be thrown away, since washing will not remove the odor. And the juice can also cause a skin irritation almost as unpleasant as poison ivy.

Because of their stately appearance, however, male ginkgo trees have been popular ornamentals since they were discovered in China's Chekiang Province. The ginkgo tree was reportedly brought to Europe from Japan in 1727, where it was planted in the Botanic Garden in Utrecht, Holland. It is thought to have been brought to the United States in 1784 and planted in John Burtram's garden in Philadelphia. The ginkgo tree is often seen in parks and gardens in London, and is widely distributed in New York City.

Ginkgo wood is soft and white and has little commercial value, although it is sometimes used for firewood and handcrafting.

Ginkgo supplements have been available over-the-counter in health food stores for several years, but the Food and Drug Administration has not yet approved for medicinal purposes the various ginkgo-derived drugs that are available in other countries.

Ginkgo has been a staple of Chinese herbal medicine for thousands of years, being recommended for coughs, asthma and acute allergic inflammations. It is one of the components of an elixir called Soma, which is a traditional Hindu medicine.

The ability to make these chemicals in the laboratory is a significant step forward, since it will help scientists to evaluate the specific health benefits of each ginkgolide compound and make it easier for drug companies to obtain large amounts of the most promising substances. At the present time, ginkgolides are obtained by extraction from ginkgo leaves, a time-consuming process that yields rather small amounts of the compounds from a large number of trees.

## Chemical Composition

Although traditional Chinese texts, such as a *A Barefoot Doctor's Manual,* refer to the medicinal properties of ginkgo seeds, seedcoats and leaves, it is the extract from the leaves that is of the greatest interest to modern-day chemists and herbalists. The *Manual* refers to a decoction made from the leaves and suggests that the seeds can be pan-fried for eating. Ginkgo is variously referred to in Chinese herbals as *kung-sun shu, fei-o-hsieh* (flying-moth leaf), *fu chi-chia* (Buddha's fingernails), *ya-chiao-pan* (duck-foot) and *lingyen.*

Extracts of ginkgo leaves are widely recommended in Asian and European medicine, and account for annual sales of about $500 million.

The leaves of the ginkgo tree are converted into a standardized preparation, such as EGb 761, GBE 24 and other formulations in Europe and the Far East. For example, EGb 761 includes 24 percent flavonoid glycosides, mostly kaempferol and quercetin glucorhamnoside esters and 6 percent of the characteristic terpenes, the ginkgolides and biloboalide. A number of other constituents make the extract water-soluble.[5]

Although the processing of the leaves is a closely guarded trade secret, the general approach is to collect the leaves in the fall, when they are turning from green to yellow. The yellow color signifies that the flavonoids are present in the leaves.

After the leaves have been harvested, they are dried in specially controlled warehouses where the air temperature remains constant. The leaves are first raked to eliminate twigs, branches, stalks and other foreign matter that might compromise the purity of the final extract. Pressed into bales and kept at a constant temperature and humidity to prevent the accumulation of moisture and fermentation, the leaves are then sent to an extraction plant.

At the plant, the leaves are pulverized and mixed with organic solvents, such as a water-acetone mixture, to extricate the chemical components in the plant. This mixture is heated and further refined to ensure purity. The procedure has been refined so that the flavonoids make up a precise 24 percent concentration, which is said to be the optimum for obtaining the therapeutic effects of ginkgo.

The flavonglycosides, which are part of the bioflavonoid family, are flavonoid molecules that are unique to ginkgo.

In addition to kaempferol and quercetin, ginkgo also contains these flavonoids: sciadopitysin, luteolin, amentoflavone, isorhamnetin, ginkgetin, delphidenon, isoginkgetin, procyanidin, bilobetin, prodelphinidin.

The chemical mixture also contains two other chemical groups, namely terpene lactones (ginkgolides A, B and C and bilobalide), and minor organic acids such as hydroxykynurenic acid, pyrocatchuic acid, kynurenic acid, vanillic acid and hydroxybenzoic acid.

Although thousands of ginkgo trees are being grown in the Far East, Europe and the United States to satisfy the growing demands for this medicinal extract, researchers at Harvard University have succeeded in synthesizing one of its most potent constituents, thereby making it even more available to researchers and scientists. No doubt this will contribute to an even greater admiration for ginkgo and its therapeutic possibilities.

# ALLERGIES

An allergy is defined as any unusual or exaggerated response to a substance, called an allergen, in someone who is sensitive to that substance. The condition used to be called hypersensitivity, but the preferred term is usually allergy.[1]

Allergies are the result of reactions of the body's immunological processes to a variety of "foreign" substances—chemical substances in foods, drugs, insect venom, etc.—or to physical conditions. The body develops immunity to these substances by forming antibodies to overcome viruses, bacteria or other substances that are not usually found in tissues. Most people are not bothered by allergies because antibodies or special mediator cells respond naturally to protein-type substances they are exposed to. But the allergy-prone person has a protective mechanism that overreacts to certain substances, which results in allergy symptoms.

Writing in *Botanical Influences on Illness: A Sourcebook of Clinical Research*, Melvyn R. Werbach, M.D. and Michael T. Murray, N.D. report that ginkgo biloba contains a number of unique terpene molecules—known collectively as ginkgolides—that antagonize platelet activating factor (PAF), a prominent chemical mediator in asthma, inflammation and allergies. PAF has a central role in numerous inflammatory and allergic processes, such as neutrophil activation, increasing vascular permeability, smooth muscle contraction, including bronchoconstriction and reduction in coronary blood flow. As reported elsewhere in this book, ginkgolides compete with PAF for binding sites and inhibit the various events caused by PAF. Mixtures of ginkgolides and ginkgo biloba extract standardized to contain 24 percent ginkgoflavonglycosides have shown clinical effects, the authors said.[2]

Researchers reported in 1990 that clinical responses to intradermal PAF were studied in 12 patients before and after they were given 120 mg/day of ginkgo biloba extract. Without the extract pretreatment, PAF produced an immediate acute wheal and flare reaction. The predominantly neutrophilic response was seen 30 minutes after the extract was given and this peaked at four hours. Eosinophils (leukocytes) were observed in the infiltrate 30 minutes after injection and remained so for 12 hours. Although the extract antagonized the acute flare response to intradermal PAF-acether, the researchers maintained this had little effect on cellular recruitment at the injection site.[3]

A daily dose of ginkgo biloba extract (240 mg/day) for three weeks brought a dramatic improvement to an adult with systemic mastocytosis, an abnormal proliferation of mast cells in some tissues, which resulted in redness of the face and trunk and recurrent facial flushing, conjunctivitis, palpitation, dizziness, abdominal pain, diarrhea, nausea and severe low blood pressure, according to a report in *Lancet* in 1988.[4]

# ALZHEIMER'S DISEASE AND MEMORY IMPAIRMENT

When it was first described by Alois Alzheimer (1864–1915), a German neurologist, in 1907, Alzheimer's disease was considered a rare disorder. The progressive, degenerative disease attacks the brain and results in impaired memory, thinking and behavior.

Today, Alzheimer's is regarded as the most common cause of dementia. However, dementia is not a disease *per se* but a group of symptoms that characterize certain diseases and conditions. In Alzheimer's, the dementia includes a decline in intellectual function that is severe enough to interfere with the ability to perform routine activities.

The second most common form of dementia is referred to as multi-infarct dementia, which is caused by vascular disease and strokes. Other causes or dementia include Huntington's disease (chorea), Parkinson's disease, Pick's disease and Creutzfeldt-Jacob disease.

A number of conditions have dementialike symptoms, such as depression, drug reactions, thyroid disorders, nutritional deficiencies, brain tumors, head injuries, alcoholism, infections (meningitis, syphilis and AIDS) and hydrocephalus (a build-up of water on the brain).

Alzheimer's develops over a period of time and there are a variety of symptoms. Typical symptoms include difficulty with memory and loss of intellectual abilities which interfere with daily work and social activities. There may be confusion, language problems, such as trouble finding the right word, poor or decreased judgment, disorientation in place and time and changes in behavior or personality. The course of the disease averages eight years from the inception of symptoms, but there have been cases lasting as long as 25 years. In severe cases, the patients are unable to care for themselves.

When studies were initiated in the 1970s, it was believed that 2.5 million Americans suffered from the disease, according to the National Institute on Aging. However, a more recent study suggests that these figures represent only the tip of the iceberg.[1]

Scientists began to revise upward the estimates of the number of Alzheimer's patients following a study reported on in the *Journal of the American Medical Association* in 1989, in which Denis Evans, M.D. and his colleagues at Brigham and Women's Hospital in Boston, Massachusetts, reported on their findings of the prevalence of dementia in a community in East Boston. The researchers found that 10.3 percent of the people over age 65 had what they called ''probable'' Alzheimer's disease.[2]

In 1982, Dr. Evans and his colleagues began studying older people living in East Boston, a typical working-class community. During that time some 3,800 residents over 65 have participated in the study. The methodology included a questionnaire concerning medical and social problems, a memory test and, for 476 volunteers, a comprehensive medical evaluation to rule out the presence of health problems other than Alzheimer's.

Dr. Evans and his team were surprised to find that the prevalence of Alzheimer's rose more rapidly with age than had previously been suspected. As an example, for the seniors between 65 and 74, 3 percent had probable Alzheimer's disease, compared to 18.7 percent in the 74- to 84-year-old age group and 47.2 percent in those over 85. This last figure for those 85 and older is almost double previous estimates.

Dr. Zaven Khachaturian, associate director for Neuroscience and Neuropsychology of Aging at the National Institute of Aging, commenting on the study, said:

> Given Dr. Evans's data, coupled with Census Bureau estimates for the numbers of people 85 and older, the actual number of Alzheimer cases in this country might be close to 4 million. Since the numbers of people over 85 are growing faster than any other segment of the U. S. population, there could be as many as 14 million Americans with Alzheimer's disease by the middle of the 21st century. Also, the prospects are very good that medical advances and changes in lifestyle will result in an even greater proportion of people living to extreme old age than the census data predicts, meaning even higher numbers of Alzheimer patients in the next century.

There are two forms of Alzheimer's disease, namely a presenile form with onset as early as age

40 and a senile form with the onset after age 60. The former is classed as familial, while the latter, much more common, is considered sporadic.

The brain, which is the main organ affected by Alzheimer's, consists of three major divisions, namely, the cerebrum, cerebellum, and brainstem. The cerebrum, or cerebral cortex, is divided into areas with sensory function, motor function and integrative function. The latter are referred to as association areas. One such area integrates information from cortical areas receiving sensations of touch, hearing and vision, while others are considered to be sites of typically human thought processes.

> Alzheimer's disease typically affects not only the cortex but also several of these areas . . . Each of these areas has different cell populations and organization, but all contain large and small neurons, glial cells, perikarya and neuropil like the central cortex.

Numerous studies have shown the efficacy of using ginkgo biloba extract to improve the mental acuity of geriatric patients. For example, W. V. Weitbrecht and W. Jansen, of Nuremberg, Germany, conducted a double-blind study involving 40 patients, ages 60 to 80, who had been diagnosed with primary degenerative dementia. During the 3-month study, one group of 20 received either Ginkgo biloba extract (120 mg/day), while the other 20 were given a placebo.[3]

The researchers reported that those receiving the ginkgo extract were more alert, scored higher on psychometric tests and had a more positive outlook than the controls. The GBE group experienced a "significant improvement," compared with no gain for the placebo group.

At the Whittington Hospital in London, researchers examined the benefits of ginkgo biloba extract on 31 patients over the age of 50 with signs of memory impairment, reported Donald J. Brown, N.D. in the May 1992 issue of *Let's Live*.[4] The study, which lasted six months, was originally published in *Current Medical Research and Opinion*.

In the double-blind study, half of the volunteers were given 40 milligrams of ginkgo biloba extract three times daily, while the other half remained on a placebo. Psychometric tests were evaluated at the beginning of the study and after 12 and 24 weeks of treatments.

The results were encouraging, Dr. Brown said:

The patients who received GBE showed significantly superior improvement compared to those given a placebo. Besides demonstrating that Ginkgo extract has a beneficial effect on mild to moderate memory loss of organic origin, the study revealed that electroencephalogram (EEG) measurements in the GBE group indicated improved brain function. This supports other research that has shown GBE increases the rate of information transmission by nerve cells.

In 1975, French researchers studied the potential benefits of ginkgo biloba extract in a group of 60 patients (55 females and 5 males), who had been diagnosed with cerebrovascular insufficiency, and 30 female patients who served as controls and were given ergot alkaloid derivatives.[5] Pretrial examinations determined the extent of dizziness, headaches, movability, sensory manifestations, etc., as well as typical psychometric tests.

Following the tests, the researchers determined that there was a 79 percent improvement in the GBE group, compared with 21 percent of the placebo. The trial group received 120 mg/day of GBE for three months.

Dr. E. W. Funfgeld reported that ginkgo biloba extract offers considerable promise in the treatment of Alzheimer's disease.[6] He added that, although there is a significant reduction in glucose consumption in patients with the degenerative type of dementia, there is a loss of cortical neurons, as well as a reduction in some subcortical structures, and a loss of the acetylcholine-synthesizing enzyme choline acetyltransferase, which causes a significant reduction in acetylcholine synthesis. Also involved are other transmitters such as serotonin.

"As reported by I. Hindmarch in a double-blind crossover study, including eight healthy female volunteers in an acute and ascending trial, one hour after 600 mg of GBE, there was a significant improvement of the short term memory which was not seen at the doses of 120 mg and 240 mg," Dr. Funfgeld said.[6,7]

He referred to a study involving a double-blind trial using GBE vs. a placebo. There were 166 patients with a mean age of 82.1 years. During the one-year study, 80 volunteers received GBE and 86 patients took a placebo. Following the study, the GBE group, which had been evaluated with the Geriatric Clinical Evaluation Scale, improved 17.1 percent, contrasted with 7.8 percent in the controls.[6,8]

In a study of his own, a 74-year-old man suffering form Parkinson's disease, memory loss, lack of energy and occasional delirious states was given infusions of 200 mg ginkgo extract daily for 10 days. Dr. Funfgeld found that the patient was more alert and had more drive following the experiment.[6,9]

"No doubt this new technique opens up big advantages in the clinical and therapeutic field. Without doubt more trials in different conditions and stages of dementias with different dosages are necessary, but, even with our present knowledge, the ginkgo biloba extract could be included in the list of therapeutics established by C. A. Bagne, et al. in 1986."[6,10]

Dr. Funfgeld and his colleagues also reported that ginkgo biloba extract appears to delay mental deterioration during the early stages of Alzheimer's disease. In fact, GBE might help to reverse some of the disabilities associated with Alzheimer's and help the patient to maintain a normal life without having to be hospitalized.[11]

# HEADACHES

In addition to being a symptom of emotional stress, headaches can be due to an infectious disease, teeth and mouth disorders, eye problems, colds and sinusitis, anemia, gastrointestinal disorders, very high blood pressure, a head injury, natural gas poisoning, a sensitivity to monosodium glutamate (MSG), among others.[1] Allergies and brain disorders are apparently only rarely causes of headaches.

In a 1975 study, GBE was given to patients suffering from migraine headaches. The results of an open trial were very good, with improvement or almost a total cure in 80 percent of the cases, the researcher said.[1,2] For patients with other types of headaches, the results were not as definitive.

The results of a double-blind study confirmed the effectiveness of using GBE to treat migraine headaches. Since these volunteers had suffered from migraines for a considerable time and had not had sufficient relief from other therapies, the researcher concluded that ginkgo biloba extract should be considered an effective supplement against migraine.

Another researcher reported in 1978 that GBE is useful in treating some patients suffering from migraines.[3]

# ASTHMA

A condition which often develops in childhood, asthma is a lung disease in which "twitchy," overactive bronchial tubes narrow, swell and become clogged with mucus. During an attack, the asthmatic has difficulty inhaling fresh air and exhaling spent air, and this often creates a wheezing or whistling sound. The patient is also likely to cough uncontrollably, have a tightness in the chest and be short of breath. Gasping for oxygen leaves the patient anxious and fatigued.

Most people with asthma are allergic to one or more substances. The most likely candidates are dust and mold, animals and birds, foods such as eggs and milk, pollen and a variety of drugs, especially those related to penicillin and related antibiotics.

Modern Chinese pharmacopoeias still list fruit and leaf extracts of ginkgo biloba as being beneficial in treating chronic bronchitis and asthma, according to noted French researcher F. V. DeFeudis.[1] In animal tests at least, bronchoconstriction has been inhibited in the presence of platelet-activating factor (PAF)-antagonistic ginkgolides in these extracts.

Writing in *Let's Live*, Donald J. Brown, N.D. discusses the ability of ginkgolides A, B and C to inhibit PAF, and notes that research regarding PAF-induced disease focuses on bronchial asthma and suggests that PAF plays a central role in the creation of long-term airway hypersensitivity and bronchial constriction noted in asthma.[2]

He went on to say that the three ginkgolides form a mixture known as BN 52063 and equals the natural ratio of the three components found in the ginkgo biloba extract. In a study published in *Prostaglandins* in 1987,[3] BN 52063 significantly inhibited bronchial constriction in asthmatic patients for up to six hours after they were administered an asthma-causing allergen. Several other studies have confirmed these results and suggest a therapeutic role for GBE in the management of asthma.

"I have had particular success with a liquid form of GBE in my pediatric asthma cases," Dr. Brown writes. "My observations have included a reduction in the severity and frequency of asthma attacks and a marked reduction in the need for bronchodilating medication."

A study in 1987 reported a double-blind trial in which eight patients who were diagnosed with bronchial asthma received 120 mg/day of ginkgo biloba extract for three days, or a placebo. After three days of treatment, the volunteers were challenged with a spray of house dust mite or pollen allergen. The researchers said that the GBE extract significantly antagonized early bronchoconstriction and showed a tendency to inhibit residual bronchial hyperreactivity, which was determined six hours after the allergen challenge by a provocation test to acetylcholine. There were no side effects reported by the participants in the study.[4]

In a discussion of platelet activating factor (PAF) in diseases of the aged, researchers reported in *Drugs & Aging* in 1992 that treatment of 17 selected asthmatics for over a year with ginkgo biloba extract resulted in an overall improvement in the management of the disease. The extract contained a mixture of the PAF receptor antagonists. Another study demonstrated that the antagonist of the ginkgo extract's group inhibited bronchoconstriction, decreased the number of hospitalizations, reduced the use of corticosteroids and improved other parameters. And oral administration of ginkgo extracts effectively inhibited bronchoconstriction induced by either inhalation of PAF, or exercise.[5]

This study utilized a mixture of ginkgolides A, B and C, which are referred to as BN 52063, and which are known to antagonize platelet activating factor, a key mediator in asthma, allergies, etc. The preparation was given orally or was inhaled by 10 patients with exercise-induced asthma.[6]

When the ginkgo mixture was given orally prior to treatment, there was no reduction in bronchoconstriction, but a significant increase of respiratory function was reported immediately after inhalation of BN 52063. The researchers also noted a significant inhibition of PAF-induced platelet aggregation after oral administration of the extract following dry cold air challenges. Oral administration of the extract was effective in improving pulmonary function as well as protecting against exercise; however, inhalation of the extract was not effective.

# TINNITUS AND HEARING LOSS

An estimated 8.5 million Americans are afflicted with hearing loss. Of this number, 71,000 may be totally deaf, while about 235,000 experience a severe hearing problem bordering on deafness. Another 8 million have a variety of hearing handicaps, with about a million American children reporting various stages of hearing impairment.

According to *The People's Medical Manual,* you should be concerned about a possible hearing problem if you: 1) hear better some days than others; 2) often fail to catch words or phrases; 3) find yourself unable to follow conversations in a group; 4) find you can better understand what a person is saying when you are facing him; 5) frequently feel that your family and friends mumble instead of speaking clearly; 6) have a running ear, or pain or irritation in the ear; 7) suffer from dizziness, loss of balance or head noises.[1]

Tinnitus or ringing in the ears is a hearing problem that can range from barely audible to quite loud, the manual continued. It affects some 9 million Americans. Most people with tinnitus report a high-pitched ringing, while others experience a buzzing, hissing, roaring or other sound. The sound can be constant or intermittent.

On the condition's causes, the *Manual* notes:

Tinnitus is often caused by exposure to loud sounds. It also can be a symptom of many conditions, accompanying virus infections, allergies and blood and circulatory disorders. Some medications—such as aspirin and quinine—may cause it. Brain cancer, meningitis and head injuries can be implicated, as can diseases of the nervous system that involve the auditory nerve. Tinnitus is usually accompanied by hearing loss.

Ginkgo biloba extract has been shown in open trials to be an effective therapeutic agent in patients with dizziness, vertigo and tinnitus. The extract, administered orally in divided doses of 60 to 160 mg/day, produced resolution or marked improvement in symptoms in between 40 and 80 percent of the volunteers who were treated, compared to those getting a placebo. The extract has been especially successful in treating patients with vestibular neuronitis, or an inflammation of inner ear nerve cells.[2]

Double-blind, placebo-controlled studies showed resolution or marked symptomatic improvement in 44 to 85 percent of patients with vertigo or dizziness treated with ginkgo biloba extract for 1 to 3 months. This is twice the usual rate of placebo response.

The causes of vertigo were typically seen to be underlying disorders such as Ménière's disease, vestibular neuropathy or infection or traumatic injury. GBE was also highly effective for patients with vertigo with no definable etiology.[2]

A significant improvement in tinnitus of less than one year's duration was reported for patients given GBE versus a placebo.[3,4] In fact, there was a distinct improvement in 50 percent of the patients within 70 and 119 days.

A 1979 study reported that GBE was successful in treating 60 patients—35 men and 25 women—with hearing loss and/or vertigo and tinnitus.[5] The problems were apparently related to various vascular disorders, aging of the inner ear, trauma or infection.

In the treatment group, the volunteers were given 120 mg/day of biloba extract, while the controls received 15 mg/day of nicergoline, a European drug. The researchers reported that GBE was effective by all criteria measured, especially with respect to vertigo and electronystagmography, a device used to record side-to-side movements of the eyeballs. The drug was also effective; however, when the dosages were compared, the extract was said to be superior.

Many studies have shown the superiority of ginkgo biloba extract in treating vertigo and tinnitus when compared with vasoregulatory drugs. However, GBE may not always be effective if the condition has lasted for many years. A number of researchers have concluded that vasoactive substances, especially GBE, are of great potential in controlling hearing loss and disturbances in equilibrium.

# IMPOTENCE

Because it is such a shock to the ego, perhaps the disorder that most men fear the most is impotence. This malfunction, which affects most men at some time in their life, is the loss of a man's ability to acquire and maintain an erection. Physicians and urologists look at both psychological and physical factors in determining a solution, however, the problem, either temporary or otherwise, can often be traced to depression, stress, fatigue, drugs and alcohol, marital discord, smoking and other factors. Impotence is not to be confused with infertility, which means that sperm are not sufficiently healthy to fertilize an egg.

During an erection, the penis becomes engorged with blood as blood vessels enlarge or dilate and allow an increased flow. This change is due to nerve stimulation, and since some nerves are ultimately controlled in brain centers, a number of drugs that affect the brain can interfere with an erection. Some drugs which can contribute to this problem are those used to treat hypertension, such as diuretics or water pills. Tranquilizers and other drugs used to treat depression can also inhibit sexual function.

Chronic alcoholism and one of its side effects, cirrhosis of the liver, can lower the amount of testosterone circulating in the bloodstream. Testosterone, the major male sex hormone, is produced in the testes and is the hormone that gives males body hair and a deeper voice, in addition to stimulating the sex drive and producing sperm.

Circulatory problems, such as arteriosclerosis, blood vessel damage resulting from diabetes, and high blood pressure may be implicated in impotence, as can chronic illness.

Problems with sexual interest and performance can also be related to diseases that inhibit the production or action of testosterone. These generally hormonal conditions include tumors of the pituitary gland or hypothalamus, which are the centers in the brain that produce and regulate hormones.

Since ginkgo biloba extract has been used successfully to treat blood pressure regulation and various vascular diseases, it should come as no surprise that GBE has a beneficial application in dealing with impotence.

A 1989 study illustrates this. Sixty patients with arterial erectile dysfunction, who had not responded to papaverine injections, the drug of choice, were treated with ginkgo biloba extract.[1] The study lasted from 12 to 18 months and some improvement was reported in six to eight weeks. The dosage was 60 mg/day. Following six months of therapy with GBE, 50 percent of the patients were able to sustain penile erections. About 45 percent of the remaining men noticed some improvement, especially after being given the supplement in conjunction with papaverine.

Papaverine (pa-PAV-er-een) is one of the vasodilators that physicians often prescribe to cause blood vessels to expand, thus increasing blood flow. It is not recommended for those with angina, glaucoma, heart disease, myocardial infarction, a recent stroke, etc.[2] It is also not recommended for Parkinson's patients, especially for those taking levodopa, and its effectiveness can be diminished by cigarette smoking.

In his practice in New York City, Robert M. Giller, M.D. prescribes a number of therapies for impotence, including ginkgo biloba, zinc, yohimbine and the vacuum constrictor device, which is available on prescription. Ginkgo helps to increase blood flow to the penis, and some patients see an improvement in about two months. In fact, he said, some men regain full potency within six months. He recommends one 40-mg capsule or tablet daily for up to six months. If there has been no improvement by then, discontinue the supplement.[3]

New applications for ginkgo biloba extracts are emerging as more is learned about the extracts' usefulness in a variety of clinical applications, especially those involving circulatory problems, according to Steven Foster in the April 1996 issue of *Better Nutrition*. As an example, he discussed a 1991 study published in the *Journal of Sex Education and Therapy* which evaluated the effect of GBE extract on erectile dysfunction in 50 patients. The volunteers, who had been diagnosed with arterial erectile impotence, were given 240 mg/day of ginkgo biloba extract for nine months.[4]

The researchers divided the men into two groups based on their response to conventional therapies. For example, 20 of the patients had previously benefited from conventional drug therapy and were selected for the first group. The second group of 30 men had remained impotent following conventional treatments.

After six months of treatment with the ginkgo bi-

loba extract, all 20 men in the first group regained
proper function, and the improvement continued
throughout the nine-month treatment period. In the
second group, 19 of the 30 volunteers responded pos-
itively to the ginkgo extract, while 11 others remained impotent. None of the participants reported
any side effects during the trials.

Foster noted that this was one of the few studies
evaluating the use of ginkgo biloba extract in the
treatment for impotence. Typical therapy for circulatory disorders is 120 mg/day and the trials usually
last from four to six weeks.

# CIRCULATORY DISORDERS

Without a constant blood supply, human beings would not be able to live. And when this constant flow of oxygen-carrying blood is disrupted, serious health problems can develop.

Blood is dispersed through the body via its central pump, the heart. Used blood is pumped to the lungs, where it picks up oxygen and discards carbon dioxide. The blood then returns to the heart and is pumped throughout the body. This supplies all of the body's tissues with nutrients and picks up waste products before returning the blood to the heart to become reoxygenated.

Your circulatory system can go awry if the central pump malfunctions or problems arise within the blood vessels. As an example, there can be a weakness in an artery wall, or the hardening of an artery that makes it unable to absorb increased blood pressure. Blood clots that cause blockages can form, and a variety of other disorders can develop. Some of these include hardening of the arteries (arteriosclerosis), deep-vein thrombosis, pulmonary embolism, thrombophlebitis, aneurysms, varicose veins, Raynaud's disease, acrocyanosis, Buerger's disease, cranial arthritis, arterial embolism, dry and wet gangrene, pulmonary hypertension, low blood pressure (hypotension), and even frostbite.

Stroke is also a possibility when part of the brain is damaged because of an impaired blood supply. Such a disturbance, characterized as cerebral thrombosis, cerebral embolism or cerebral hemorrhage, can result in a deterioration of both physical and mental acuity. Cerebral thrombosis is usually due to the narrowing of an artery that supplies blood to the brain, a complication of hardening of the arteries. A cerebral embolism is caused by a foreign object, or embolus, which is carried in the bloodstream and becomes wedged in a place where it inhibits blood flow to the brain. A cerebral hemorrhage simply means that an artery bursts.

A 1965 study reported that ginkgo biloba extract lowered blood pressure and dilated or expanded the peripheral blood vessels, including capillaries, in 10 patients with postthrombotic syndrome.[1] This did not increase capillary permeability but it did reduce the swelling.

In 1972 the use of ginkgo biloba extract was compared with other vasodilators, notably hydrogenated alkaloids of ergot, acetylcholine chloride and sodium nicotinate.[2] All patients had varying degrees of vascular disease. The research team stated that the vasodilator action of GBE is similar to the other substances but is significantly more constant.

A study in 1977 was conducted to determine the activity of GBE on cerebral blood flow in 20 patients, ages 62 to 85, who were diagnosed with cerebral circulatory insufficiency, due to age and hardening of the arteries.[3] The patients were treated orally and intramuscularly for 15 days. Because of the age and health of the volunteers, the researchers maintained low dosages of GBE and did not expect spectacular results. However, they reported that the cerebral hemodynamics was much improved in 15 of the cases.

In another 1977 study[4,5] researchers reported functional improvement in 65 percent of patients with arterial leg disease following GBE therapy. There was only a 22.5 percent response in the placebo group. Some patients in the treatment group also reported a resolution of trophic skin conditions, better circulation to the extremities and a lessening of impotence, among other things.

About two-thirds of the patients receiving GBE showed definite clinical improvement, compared with 16 percent given a placebo, when treated for a variety of peripheral vascular diseases.[4,6] After further analysis of the 1975 study, the authors reported that the supplement was 100 percent effective in patients with Grade II lower limb arteritis (inflammation of an artery). There was a 33 percent response in volunteers with Raynaud's disease (which involves fingers and toes), but there was no response in those with Grade III arteritis.

The authors went on to say that, when compared with the placebo group, GBE produced much higher rates of improvement of peripheral pain (66 vs 13 percent); intermittent claudication (64 vs 19 percent); warmth of lower limbs (64 vs 19 percent); clearing of ulcerous lesions (100 vs 0 percent); trophic changes in lesions (100 vs 25 percent); and in pain attacks in Raynaud's disease (33 vs 0 percent).

In an open comparison, GBE—160 mg/day given orally for six months—was more effective than buflomedil (600 mg/day) in 38 patients with peripheral occlusive arterial disease and was equivalent to pentoxifylline (1,200 mg/day) in another 27 patients participating in a double-blind study with respect to improving walking distance, relieving pain and increasing microcirculation.[4,7,8]

In 1984, a double-blind study tested the efficacy of GBE in two groups of patients with peripheral arterial insufficiency.[9] The study lasted six months. There was a marked improvement in walking without pain and increased blood flow to lower limbs in the treatment group. The author of the report stated that the improvement rate was not only statistically significant but clinically remarkable.

Following success with GBE in smaller trials, researchers in 1967 conducted a larger study involving patients with Parkinson's disease secondary to cerebral arteriosclerosis.[10] The supplement was given either orally or by intravenous injection.

When compared with standard vasodilator therapy in 40 postsurgical patients with lower limb arterial obstruction, GBE, in another study, brought improvement in resting and walking pain. The dosage was 160 mg/day.[4,11]

Twenty-one patients with chronic arteriopathies of the lower extremities were treated with GBE in an experiment going back to 1975. The volunteers, with a mean age of 60 years, received 160 mg/day of GBE for a month, and the trial provided encouraging results.[12,13] Functional symptoms of intermittent claudication were markedly decreased and all subjects exhibited excellent tolerance to the therapy, the researchers reported.

GBE enhances blood flow not only through the large blood vessels, but also small vessels like the

capillaries close to the skin. In 1992 Donald J.
Brown, N.D. reported in *Let's Live* that a study at
the University of Saarland in Hamburg, Germany,
revealed that GBE increases microcirculation through
the capillaries of the body. The study involved 10
patients whose skin microcirculation was monitored
every 30 minutes. The results showed a 57 percent
increase in blood flow through the nail fold capillaries of the finger after one hour.[14,15]

# EYE DISORDERS

One of the leading causes of blindness in people over 65, macular degeneration is of unknown etiology and usually develops gradually. This eye disorder involves the macula, the area of the retina near the optic nerve at the back of the eye. The macula is responsible for fine reading vision at the center of the field of vision.

Antioxidants such as vitamin A and vitamin C are free-radical scavengers, and have been studied as a possible deterrent to macular degeneration.[1] Since ginkgo biloba extract also serves as an antioxidant and a dispersant of free-radicals, this supplement has also been investigated for the treatment of macular degeneration, although on a rather modest scale.

In 1986 French researchers administered GBE, 80 mg twice daily, or a placebo, to 20 elderly patients with recently diagnosed macular degeneration[2,3] The randomized, double-blind study lasted for six months. Aside from hardening of the arteries, the seniors apparently had few other complaints.

At study completion, funduscopic examination revealed that distant visual acuity [in the most affected eye] had improved by 2.3 diopters [a measurement of refraction power] in Ginkgo biloba recipients, whereas in placebo patients the mean increase was only 0.6 diopters. Definite clinical improvements were demonstratable in 9 of 10 GBE recipients versus 2 of 10 placebo patients and, as expected, ginkgo biloba extract was significantly more effective overall. . . . This therapeutic trial, although including a limited number of case histories, seems to us to furnish proof of the value of GBE in the treatment of recent senile macular degeneration. In fact, after six months of treatment, a significant improvement was found in acuity of distance vision, which for the patient is obviously an essential criterion in his life relationships.

In experimental studies, especially with laboratory animals, GBE has proved to be an effective deterrent to free-radical damage to the retina of the eye. At least one study proved that GBE prevents diabetic retinopathy in alloxan-induced diabetic rats.[4] This suggests that ginkgo biloba extract might also be useful in treating human beings with this disorder.

In a double-blind study involving 20 patients with senile macular degeneration, 10 patients were given 80 mg of Ginkgo biloba extract morning and evening, while the remaining 10 volunteers received a placebo. The study lasted for six months. All of the patients were over 55 years of age, and their senile maculopathy had been diagnosed for less than one year.[5]

The research team concluded that, although the trial involved a limited number of participants, this was sufficient to convince the team that ginkgo is a valuable treatment for recent senile macular degeneration.

In the *Journal of the Advancement of Medicine* in 1993, Alan R. Gaby, M.D. and Jonathan V. Wright, M.D. extensively reviewed the literature concerning specific nutrients that may help cataracts and macular degeneration, which are the main causes of blindness in the United States. Nutrients which may help these conditions, they noted, include ginkgo biloba, zinc, taurine, vitamin A, vitamin B2 (riboflavin), vitamin C, vitamin E, selenium, beta-carotene (provitamin A), N-acetylcysteine and flavonoids. In their own practices, the physicians have treated approximately 60 patients with macular degeneration using intravenous zinc, selenium and other minerals.[6]

Chronic cerebral retinal insufficiency syndrome is a common complaint among the elderly. To determine the efficacy of using ginkgo biloba extract for this condition, a two-part trial was conducted involving 24 volunteers—four men and 20 women—whose age was 74.9, plus or minus 6.9 years. At the outset, retinal blood flow measurements were recorded. Using a randomized, double-blind study in two phases with two doses of ginkgo biloba extract (EGb 761), the researchers studied the effect of the extract on the reversibility of visual field disturbances. The main purpose of the study was to investigate the change in the luminous density difference threshold following the GBE therapy.[7]

In the first group of patients, the researchers reported a significant increase in retinal sensitivity following the administration of 160 mg/day of GBE for four weeks. In the second group of volunteers, who were given 80 mg/day of GBE, the change in retinal sensitivity was not determined until after the patients had received 160 mg/day of GBE. Following the trial, both doctors and patients stated that there was a considerable improvement in their general condition after the course of therapy.

# DIABETES

Researchers conducting an animal experimental study reported in 1986 that they produced diabetes in rats by injecting them with alloxan. Compared to the untreated diabetic animals, the electroretinograms of the rats treated with ginkgo biloba extract had a significantly greater amplitude after two months, suggesting that the extract may protect against diabetic retinopathy. The free radical scavenging properties of the extract are thought to be the reason for this protection.[1] (See the section on Eye Disorders in this book).

A 1992 study indicated that patients with diabetic neuropathy may be helped with ginkgo biloba extract. For example, 40 patients with this condition showed a significant increase in nerve conductivity after 14 days of therapy after receiving 100 mg/day of the extract and 15 mg/day of folic acid, the B vitamin, compared with volunteers getting placebo with or without the vitamin. There was significant improvement after an additional 14 days when the patients were given 60 mg of the extract three times a day and 5 mg of folic acid once a day. The researchers reported that the ginkgo biloba extract and the B vitamin also reduced the severity of pain, paraesthesias and dysthesias.[2]

In another study involving 10 patients with polyneuropathies as a result of diabetes, lead intoxication and abnormal amounts of protein in the blood or peripheral arterial occlusion, autonomic nervous regulation improved considerably after the patients were given intravenous doses of 87.5 mg of ginkgo biloba extract and 3 mg of folic acid. The therapy lasted for 14 days. Afterwards, subjective sensation had improved in seven patients, while six patients reported an improvement in depth sensitivity.[3]

# MULTIPLE SCLEROSIS

As reported elsewhere in this book, ginkgo biloba extracts contain ginkgolides which antagonize platelet activating factor (PAF), a key mediator in several allergic and inflammatory processes such as multiple sclerosis.

For example, a 1992 study reported that 10 volunteers with relapsing-remitting MS in acute relapse were given a five-day course of ginkgolide B intravenously. It was found that eight patients had improvement in their neurological score beginning two days after the therapy began. The improvement was sustained in five patients six days afterward. Some of the patients experienced mild side effects, but none were considered serious.[1]

In his definitive book *Ginkgo Biloba Extract (EGb:761): Pharmacological Activities and Clinical Applications*, F. V. DeFeudis reported that bilobalide—a constituent of the terpenoid fraction of EGb 761—might provide therapy for neurological disorders that are caused by, or are associated with, pathological changes in the myelin sheaths of nerve fibers.[2] Damage to myelin can occur directly, the primary form, or indirectly, the secondary form. The primary form would include inflammatory and immunological demyelinating diseases, such as multiple sclerosis and postinfectious encephalitis, whereas the secondary form would include traumatic and sclerosing neuropathies and diabetic- or alcohol-related hereditary and vascular polyneuropathies.[3]

In the Whitaker Wellness Program, as outlined in *Dr. Whitaker's Guide to Natural Healing,* Julian Whitaker, M.D. said that antioxidants that he recommends are of great importance for patients with multiple sclerosis. MS is characterized by increased lipid peroxidation of nerve membranes. In addition, whenever the level of polyunsaturated fats is increased, so is the need for vitamin E, selenium and other antioxidants.[4]

Whitaker also recommends ginkgo biloba extract—standardized to contain 24 percent ginkgo flavongycosides—at a dose of 40 mg three times daily. Ginkgo is a potent antioxidant and also improves nerve function, he added.

He also recommended omega-3 oils because of their greater effect on platelets and the requirement for these oils in normal myelin integrity. His therapy is similar to that of Dr. Roy Swank, professor of neurology at the University of Oregon Medical School in Portland, which includes the liberal consumption of fish and supplementation with cod liver oil, a rich source of omega-3 polyunsaturated oils. Whitaker also recommends two tablespoons of flax seed oil daily for MS patients.

# BRAIN TRAUMA

A 1991 study indicated that ginkgo biloba extract has a potential use in brain-injured patients, since the extract has shown little toxicity in animals and man. Their study evaluated the use of the extract in two animal models with cortical hemiplegia. In all of the animals with motor impairment, which were evaluated for seven to 30 days, there was a more rapid recovery in the rats given GBE versus the control animals getting a saccharin solution.[1]

Writing in *Ginkgo Biloba Extract (EGb 761): Pharmacological Activities and Clinical Applications*, F. V. DeFeudis stated that a number of studies have shown the value of EGb in dealing with head injuries. As an example, a 1989 study[2] demonstrated that long-term treatment with ginkgo biloba extract facilitated recovery from penetrating traumatic injuries to the cerebral cortex in rats. the anti-edema effect of EGb may be responsible for the success in this application.[3]

In his book *Herbal Prescriptions for Better Health*, Donald J. Brown, N.D. reported that he now recommends 240 mg/day of ginkgo biloba extract for patients who have experienced head injury or trauma. He came to this conclusion after treating a 36-year-old woman who had had sleep disturbances, dizziness while exercising, headaches and short-term memory loss for nine months following a blow to her head, which was caused by a fall in her kitchen. Her symptoms included frequent disorientation, especially while driving.[4]

Brown started her with 120 mg/day, in three divided doses, of GBE. She called 10 days later to say that a previously prescribed antidepressant had been stopped and that she was feeling better. A four-week follow-up showed significant improvement and that she was sleeping well. There was more mental clarity, improved short-term memory, and she was able to exercise for 20 to 30 minutes a day without dizziness or discomfort. She had returned to work part time. At an eight-week follow-up, the patient was now working full time and she did not report any of the problems outlined during her initial visit. After nine months, she was able to discontinue using the ginkgo biloba extract. Two years later she was living a normal life with no work- or activity-related problems.

# FREE-RADICAL SCAVENGER

One of ginkgo biloba extract's most notable attributes is its antioxidant, free-radical scavenging properties. Like vitamin A, beta carotene, vitamin C, vitamin E, selenium, etc., GBE helps to purge the body of potentially damaging free radicals. One of the reasons is that GBE contains such bioflavonoids as quercetin, kaempferol and rutin—which are related to vitamin C—some vitamin C and a variety of other antioxidants such as superoxide dismutase (SOD).

Free radicals are highly unstable, highly reactive molecules or fragments of molecules characterized by an unpaired free electron that is avid to grab almost anything it can grab.

In *The Complete Guide to Anti-Aging Nutrients*, Sheldon Saul Hendler, M.D., Ph.D., says:

> Free radicals are typically toxic oxygen molecules that severely damage most of the molecules they grab hold of (cell membranes and fat molecules are favorite targets). It is one of the fundamental ironies of life that oxygen both sustains us and kills us. We often forget how toxic oxygen is, though we need only look around us to be reminded of it. Most of the rust and decay that we encounter is due to oxidation. Much of the "rust and decay" of the human body is due to the same thing.[1]

Dr. Hendler also notes that to protect us from the toxic forms of oxygen—such as superoxide, singlet oxygen, hydroxy radicals—which spin off as a result of various metabolic processes, plus others which enter the body via food and air pollution, or which are byproducts of radiation, viruses and the like, we depend on free-radical scavengers.

Some of these scavengers, such as superoxide dismutase and glutathione peroxidase, are enzymes that neutralize free radicals. Other scavengers include vitamin C and selenium.

If the free-radical scavengers are not present in sufficient quantity and at the right places at the right times, the free radicals can do considerable damage to cells and to the genetic program itself. A certain amount of this kind of damage occurs almost all the time in each of us. Free radicals also help promote cross-linking and may help create destructive and sometimes malignant mutations. In polyunsaturated fats, free radicals help produce aldehydes and other substances that may produce cancer and other damage. The free-radical theory of aging is probably the most useful theory we have at the present time—from the standpoint of finding practical means of delaying the effects of aging.[1]

The human body is especially vulnerable to free-radical attacks during ischemia or a lack of blood to specific organs, stated *Ginkgo Biloba Extract (EGb 761) in Perspective.*[2] With an overabundance of free radicals, the defense mechanisms are unable to nullify these foreign particles and peroxidation of membrane fats occurs and the damage follows. GBE, especially the flavonoid glycoside constituent, destroys these excess free radicals, the publication added.

In a series of in vitro studies, a solution of 500

mg/dl of GBE was shown to inhibit formation of the hydroxyl radical by 65 percent and adriamycyl radical generation by 50 percent. A 1988 study noted that GBE also halted lipid oxidation.[2,3]

In vitro and in vivo studies in France reported that GBE, because of its content of quercetin and kaempferol esters, is a potent free-radical scavenger.[4] In two animal studies, GBE had little effect on cardiac functional parameters, but it induced a significant decrease in the intensity of ventricular fibrillation. For human hearts, however, GBE provided effective protection against the electrocardiographic disorders induced by ischemia. For other types of diminished blood supply, the researchers noted that GBE brought a decrease in arrhythmia without any change in cardiovascular parameters.

Since a number of the in vitro and in vivo experiments have shown that GBE possesses antioxidant properties, the extract can theoretically offset many of the problems associated with excessive free-radical information, according to F. V. DeFeudis.[5] This would include ischemia, problems associated with biological aging of tissues and those disease states that are related to ischemia and accelerated biological aging.

It is also of interest that [researchers] have purified and characterized an iron-containing SOD from ginkgo biloba leaves. Such iron-containing SODs are found in only a few phylogenetically diverse higher plants. . . . They could possibly contribute to the antioxidant activity of intravenously-injected GBE if present in sufficient amount in this extract.[6]

There is increasing evidence that most diseases, at some point in their development, are related to tissue injury caused by free-radical reactions, according to

U. Stein, New York Institute for Medical Research, in the August 1994 issue of *Revista Brasileira de Neurologia*. In fact, chronic or resistant diseases for which successful treatments have not been found are said to involve free-radical damage.[7]

Although a "free-radical disease" *per se* does not exist, Stein said that the following disorders are initiated or worsened by free-radical reactions: alcoholism, Alzheimer's disease, brain edema, dementia, demyelination (the destruction of myelin, which forms a sheath around nerve fibers), head trauma, Parkinson's disease, retinal damage, stroke, spinal cord damage, aging, inflammation, hardening of the arteries, intestinal ischemia, shock, transplantations, pancreatitis, smoking and myocardial ischemia.

"EGb 761 has been shown experimentally to affect the organism at the organ, tissue and cell level in various systems," Stein observed. "*In vitro* and *in vivo* EGB 761 seems to protect tissue against metabolic stress, which can be taken as indirect proof for its antioxidant effectiveness. The protection of retinal structures against laser irradiation in rabbits, the protection of astrocytes [star-shaped cells] against toxic chemicals in rats and the reduction of brain edema after frontal lobe trauma in rats stand exemplary for EGB's mode of action. All of this might be reduced to the protection of membranes against lipid peroxidation."

An extract of ginkgo biloba (100 mg/kg/day) decreased damage to retinal blood vessels induced by oxygen free radicals in a 1990 study, and protected retinal cells from damage induced by photocoagulation.[8]

# DOSAGE

In prescribing ginkgo biloba extract for various health problems, researchers generally recommend a dosage of 40 mg three times daily. This is given as drops of a standardized GBE containing 40 mg/dl of extract, including 9.6 mg of flavonoid glycosides. Others have reported that the standardized extract contains 24 percent flavoglycosides.[1,2]

As reported in this book, other researchers have used higher dosages of ginkgo biloba extract or GBE tablets. A slightly higher dosage of GBE is listed at 160 mg/day. This dosage, recommended for vertigo, tinnitus and peripheral vascular disease, can be given in divided doses, such as 80 mg twice daily or 40 mg four times each day. For these complaints, researchers have also suggested 20 to 80 mg of GBE given three times a day.

When not using the standardized GBE, it is apparently difficult to devise a dosage, since there may be extreme variations in the active constituents of the dried leaves and crude extracts. In any case, a standardized extract for content and activity is recommended by researchers.

# POTENTIAL
# SIDE EFFECTS

There have been few side effects from taking ginkgo biloba extract reported in the scientific literature. Higher dosages have sometimes produced complaints, perhaps because of the presence of various constituents that are generally removed or altered by the standard extraction process. The usual complaints include a mild gastrointestinal upset or headache. However, some people have reported severe allergic reactions from the ginkgo fruit pulp.[1]

In a 1988 study, only 33 of 8,505 patients receiving GBE experienced side effects. Nine of these patients did complain of gastrointestinal upsets.[2,3]

In other studies, several patients experienced mild nausea and heartburn.[2,4,6] And two patients given GBE suffered severe nausea and vomiting.[2]

Obviously, patients taking GBE for a mild to serious health problem should have their progress monitored by a physician or other professional. However, as a therapeutic aid for keeping well, many people can benefit from the daily use of over-the-counter preparations.

# Conclusion

Few herbal remedies have been as extensively researched as has ginkgo biloba extract, and the supplement seems to be one of our most useful therapies for many complaints. As reported throughout this book, GBE is an exceptional therapeutic agent for the treatment of some patients with Alzheimer's disease, asthma, impotence, tinnitus, migraine headaches, strokes, hemorrhoids, various circulatory disorders, depression and many other illnesses.

Since GBE is so compatible with other medications, it could prove to be useful in combination therapy, according to Francis V. DeFeudis[1] As an example, 10 patients, ranging in age from 30 to 70, who suffered from painful diabetic neuropathy, experienced a significant decrease in pain on the fifth and tenth days of the study, after receiving GBE and folic acid, the B vitamin. GBE was prescribed 87.5 mg/day and the folic acid dosage was 3 mg/day.

Extensive clinical trials do not seem to be necessary to guarantee the safety of GBE-containing products, DeFeudis says, since its efficacy has been well documented. And even if individual constituents in GBE continue to be analyzed, it is doubtful that they will be as significant as the total extract.

# REFERENCES

**ALLERGIES**

1. Ensminger, A., et al. *Foods and Nutrition Encyclopedia.* Clovis, Calif.: Pegus Press, 1983, pp. 42ff.

2. Werbach, Melvyn, M.D., and Murray, Michael T., N.D. *Botanical Influences on Illness.* Tarzana, Calif.: Third Line Press, 1994, p. 48.

3. Markey, A. C., et al. "Platelet Activating Factor-Induced Clinical and Histopathologic Responses in Atopic Skin and Their Modification by the Platelet Activating Factor Antagonist BN52063," *J. Am. Acad. Dermatol.* 23(2):263–268, 1990.

4. Guinot, P., et al. "Treatment of Adult Systemic Mastocytosis with a PAF-Acether Antagonist BN52063," *Lancet* ii:114, 1988.

**ALZHEIMER'S DISEASE AND MEMORY IMPAIRMENT**

1. Emr, Marian. "Scientists Revise Estimates on Prevalence of Alzheimer's Disease," *National Institute on Aging Notes,* November 9, 1989, unpagenated.

2. Evans, D. A., et al. "Clinically-Diagnosed Alzheimer's Disease: An Epidemiologic Study in a Community Population of Older Persons," *Journal of the American Medical Association,* November 10, 1989.

3. Weitbrecht, W. V., and Jansen, W. "Double-blind and Comparative (Ginkgo Biloba vs. Placebo) Therapeutic Study in Geriatric Patients with Primary Degenerative Dementia—a Preliminary Evaluation." In *Effects of Ginkgo Biloba Extract on Organic Cerebral Impairment,* A. Agnoli, et al. London: Eurotext Ltd., 1985, p. 91–99.

4. Brown, Donald J., N.D. "Ginkgo Biloba—Old and New: Part II," *Let's Live,* May 1992, pp. 62–64.

5. Moreau, P. "Un Nouveau Stimulant Circulatoire Cerebral," *Nouv. Presse Med.* 4:2401–2402, 1975.

6. Funfgeld, E. W., editor *Rokan (Ginkgo Biloba): Recent Results in Pharmacology and Clinic.* Berlin: Springer-Verlag, 1988, pp. 11–12ff; p. 49–54; p. 99; pp. 278–286.

7. Hindmarch, I. "Activite de Ginkgo Biloba sur la Memoire a Court Terme," *Presse Med.* 15:1592–1594, 1986.

8. Taillandier, J., et al. "Traitement des Troubles du Vieillissement Cerebral par L'Extrait de Ginkgo Biloba, Etude Longitudinale Multicentrique a Double Insu Face au Placebo," *Presse Med.* 15:1583–1587, 1986.

9. Funfgeld, E. W., and Stalleicken, D. "Dynamic-Brain Mapping," *TW Neurol. Psychiatr.* 2:136–142, 1987.

10. Bagne, C. A., et al. "Alzheimer's Disease: Strategies for Treatment and Research." In "Treatment Development Strategies for Alzheimer's Disease," F. Crook, et al., editors. Madison, Conn.: Mark Pawley Assn., 1986, pp. 585–636.

11. Funfgeld, E. W. "A Natural and Broad Spectrum Nootropic Substance for Treatment of SDAT—the Ginkgo Biloba Extract." In "Alzheimer's Disease and Related Disorders," Iqbak, K., et al., editors. New York: Alan Liss, 1989, pp. 1247–1260.

## HEADACHES

1. DeFeudis, F. V. *Ginkgo Biloba Extract (EGb 761): Pharmacological Activities and Clinical Applications.* Paris: Elsevier, 1991, p. 142.

2. Dalet, R. "Essai du Tanakan dans les Cephalees et les Migraines," *Extr. Vie Med.* 35:2971–2973, 1975.

3. Devic, M. "Le Tanakan dans le Traitement de Fond de la Migraine," *Lyon Mediterr. Med.* 239:735–738, 1978.

## ASTHMA

1. De Feudis, F. V. *Ginkgo Biloba Extract (EGb 761): Pharmacological Activities and Clinical Applications.* Paris: Elsevier, 1991, p. 92.

2. Brown, Donald J., N. D. "Ginkgo Biloba—Old and New: Part II," *Let's Live*, May 1992, pp. 62–64.

3. *Prostaglandins* 34(5), 1987.

4. Guinot, P., et al. "Effect of BN 52063, a Specific PAF-Acether Antagonist, on Bronchial Provocation Test to Allergens in Asthmatic Patients. A Preliminary Study," *Prostaglandins* 34(5):723–731, 1987.

5. Kroegel, Claus, et al. "The Pathophysioogical Role and Therapeutic Implications of Platelet Activating Factor in Diseases of Aging," *Drugs & Aging* 2(4):345–355, 1992.

6. Wilkens, J. H., et al. "Effects of a PAF-Antagonist (BN 52063) on Bronchoconstriction and Platelet Activation During Exercise Induced Asthma," *British Journal of Clinical Pharmacology* 29(1):85–91, 1990.

## TINNITUS AND HEARING LOSS

1. Lewis, Howard R., and Martha E. *The People's Medical Manual.* Garden City, N. Y.: Doubleday & Co., Inc., 1986, p. 535; pp. 297–299.

2. Chesseboeuf, L., et al. "Comparative Study of Two Vasoregulators in Syndromes of Deafness and Vertigo," *Medicine du Nord et de l'Est* 5:534, 1979.

3. *Ginkgo Biloba Extract (EGb 761) in Perspective.* Auckland, New Zealand: ADIS Press Limited, 1990, pp. 11ff.

4. Meyer, B. "A Multicenter Randomized Double-Blind Study of Ginkgo Biloba Extract Versus Placebo in the Treatment of Tinnitus," *Presse Med.* 15:1562–1564, 1986.

5. DeFeudis, F. V. *Ginkgo Biloba Extract (EGb 761): Pharmacological Activities and Clinical Applications.* Paris: Elsevier, 1991, pp. 11ff.

## IMPOTENCE

1. Sikora, R. et al. "Ginkgo Biloba Extract in the Treatment of Erectile Dysfunction," *Journal of Urology* 141:188A, 1989.

2. *The Complete Drug Reference.* Mount Vernon, N. Y.: Consumer Reports Books, 1991, p. 945.

3. Giller, Robert M., M.D., and Matthews, Kathy. *Natural Prescriptions.* New York: Ballantine Books, 1994, pp. 210–211.

4. Foster, Steven. "Ginkgo Biloba: A Living Fossil for Today's Health Needs," *Better Nutrition*, April 1996, p. 59.

# CIRCULATORY DISORDERS

1. Trommier, H., "Klinisch—Pharmakologische Untersuchungen ueber den Effect eines Extraktes aus G. Biloba L. beim post Thrombotischen Syndrom," *Arzneim. Forsch.* 18:551, 1968.

2. Gautherie, M., et al. "Effet Vasodilatateur de l'Extrait de Ginkgo Biloba Mesure par Thermometrie et Thermographie Cutanees," *Therapie* 27:881, 1972.

3. Safi, N., and Galley, P. "Tanakan et Cerveau Senile. Etude Radiocirculographique," *Bordeaux Medical* 10:171–176, 1977.

4. *Ginkgo Biloba Extract (EGb 761) in Perspective.* Auckland, New Zealand: ADIS Press Limited, 1990, p. 14ff.

5. Courbier, R., et al. "Double-Blind, Cross-Over Study of tanakin in Arterial Diseases of The Legs," *Mediterranee Medicale* 126:61–64, 1977.

6. Frileux, C., and Cope, R. "The Concentrated Extract of Ginkgo Biloba in Peripheral Vascular Disease," *Cahiers d' Arteriologie de Royal* 3:117–122, 1975.

7. Berndt, E. D., and Kramar, M. "Drug Treatment of Peripheral Arterial Occlusive Disease in Stage IIB," *Therapiewoche* 37:2815–2819, 1988.

8. Bohmer, D., et al. "The Treatment of PAOD (Peripheral Arterial Occlusive Disease) with Ginkgo Biloba Extract (GBE) for Pentoxyfyline," *Herz/Kreislauf* 20:5–8, 1988.

9. Bauer, U. "6–Month Double-Blind Randomised Clinical Trial of Ginkgo Biloba Extract Versus Placebo in Two Parallel Groups in Patients Suffering from Peripheral Arterial Insufficiency," *Arzneim. Forsch.* 34:716, 1984.

10. Hemmer, R., and Tzavellas, O. "Zur Zerebralen Wirksamkeit eines Pflanzenpraparates aus Ginkgo Biloba," *Arzneim. Forsch.* 17:491, 1967.

11. Bastide, G., and Montsarrat, M. "Arterite des Membres Inferieurs. Interet du Traitement Medical apres Intervention Chirurgicale. Analyse Factorielle," *Gaz. Med. (France)* 85:4523–4526, 1978.

12. DeFeudis, F. V. *Ginkgo Biloba Extract (EGb 761): Pharmacological Activities and Clinical Applications.* Paris: Elsevier, 1991, p. 117ff.

13. Ambrosi, C., and Bourde, C. "New Medical Treatment for Arterial Disease of the Lower Extremities: Tanakan. Clinical Trial and Liquid Crystal Study," *Gazette Medicale de France* 82(6):628–633, 1975.

14. Brown, Donald J., N. D. "Ginkgo Biloba—Old and New: Part II," *Let's Live,* May 1992., pp. 62–64.

15. *Arzneim.-Forsch. Drug Research,* Vol. 40, No. 5, 1990.

## EYE DISORDERS

1. Goldberg, Jack, et al. "Factors Associated with Age-Related Macular Degeneration," *American Journal of Epidemiology* 128(4):700–710, 1988.

2. *Ginkgo Biloba Extract (EGb 761) in Perspective.* Auckland, New Zealand: ADIS Press Limited, 1990, p. 17.

3. Lebuisson, D. A., et al. "Treatment of Senile Macular Degeneration with Ginkgo Biloba Extract: A Preliminary Double-Blind Study Versus Placebo," *Presse Med.* 15:1556–1558, 1986.

4. Doly, M. "Effect of Ginkgo Biloba Extract on the Electrophysiology of the Isolated Diabetic Rat Retina," *Presse Med.* 15:1480–1483, 1986.

5. Lebuisson, D.A., et al. "Treatment of Senile Macular Degeneration with Ginkgo Biloba Extract," *Rokan (Ginkgo Biloba). Recent Results in Pharmacology and Clinic.* Berln: Springer-Verlag, 1988, pp. 231–236.

6. Gaby, Alan R., M.D., and Wright, Jonathan V., M.D. "Nutritional Factors in Degenerative Eye Disorders: Cataract and Macular Degeneration," Journal of *The Advancement of Medicine* 6(1):27–40, Spring 1993.

7. Raabe, A., et al. "Therapeutic Follow-Up Using Automatic Perimetry in Chronic Cerebroretinal Ischemia in Elderly Patients. Prospective Double-Blind Study with Graduated Dose of Ginkgo Biloba Treatment (EGb 761)," *Klin Monatsbl Augenheilkd* 199(6):432–438, 1991.

## DIABETES

1. Doly, M., et al. "Effect of Ginkgo Biloba Extract on the Electrophysiology of the Isolated Retina from a Diabetic Rat," *Presse Med.* 15(31):1480–1483, 1986.

2. Koltringer, P., et al. "Ginkgo Biloba Special Extract EGb 761 and Folic Acid in Diabetic Neuropathia, a Randomized, Placebo-Controlled, Double-Blind Study," *Z. Allg. Med.* 68:69–102, 1992.

3. Koltringer, P., et al. "Ginkgo Biloba Extract and Folic Acid in the Treatment of Autonomic Neuropathies," *Acta Medica Austriaca* 16:35–37, 1989.

## MULTIPLE SCLEROSIS

1. Brochet, B., et al. "Pilot Study of Ginkgolide B, a PAF-Acether Specific Inhibitor in the Treatment of Acute Outbreaks of Multiple Sclerosis," *J. Rev. Neurol.* 148(4):299–301, 1992.

2. Chatterjee, S.S., et al. "Pharmaceutical Compositions Containing Bilobalid for the Treatment of Neuropathies." U.S. Patent No. 4,571,407, February 18, 1986.

3. DeFeudis, F. V. *Ginkgo Biloba Extract (EGb 761): Pharmacological Activities and Clinical Applications.* Paris: Elsevier, 1991, pp. 74–75.

4. Whitaker, Julian, M.D. *Dr. Whitaker's Guide to Natural Healing.* Rocklin, Calif.: Prima Publishing, 1995, p. 309.

## BRAIN TRAUMA

1. Brailowski, S., et al. "Effects of a Ginkgo Biloba Extract in Two Models of Cortical Hemiplegia in Rats," *Restorative Neurology and Neuroscience* 3:267–274, 1991.

2. Attella, M. J., et al. "Ginkgo Biloba Extract Facilitates Recovery from Penetrating Brain Injury in Adult Male Rats," *Exp. Neurol.* 105:62–71, 1989.

3. DeFeudis, F. V. *Ginkgo Biloba Extract (EGb 761): Pharmacological Activities and Clinical Applications.* Paris: Elsevier, 1991, pp. 141–142.

4. Brown, Donald J., N.D. *Herbal Prescriptions for Better Health.* Rocklin, Calif.: Prima Publishing, 1996, p. 127.

## FREE-RADICAL SCAVENGER

1. Hendler, Sheldon Saul, M.D., Ph.D. *The Complete Guide to Anti-Aging Nutrients.* New York: Simon and Schuster, 1985, pp. 32–33.

2. *Ginkgo Biloba Extract (EGb 761) in Perspective.* Auckland, New Zealand: ADIS Press Limited, 1990, p. 3.

3. Pincemail, J., and Deby, C. "The Antiradical Properties of Ginkgo Biloba Extract." In *Rokan (Ginkgo Biloba): Recent Results in Pharmacology and Clinic,* F. W. Funfgeld, editor. Berlin: Springer-Verlag, 1988, pp. 71–82.

4. Guillon, J. M., et al. "Effects of Ginkgo Biloba Extract on Two Models of Experimental Myocardial Ischemia." In *Rokan (Ginkgo Biloba),* p. 153.

5. DeFeudis, F. V. *Ginkgo Biloba Extract (EGb 761): Pharmacological Activities and Clinical Applications.* Paris: Elsevier, 1991, p. 51ff.

6. Duke, M. V., and Salni, M. L. "Purification and Characterization of an Iron-Containing Superoxide Dismutase from a Eukaryote, Ginkgo Biloba," *Arch. Biochem. Biophys.* 243:305–314, 1985.

7. Stein, U. "Free Radicals and Antioxidants," *Revista Brasileira de Neurologia* 30 (Suppl 1):12S–17S, August 1994.

8. Chrisp, Paul, et al. *Ginkgo Biloba Extract (EGb 761) in Perspective.* Auckland, New Zealand: ADIS International, 1993, pp. 3–4.

## DOSAGE

1. *Ginkgo Biloba Extract (EGb 761) in Perspective.* Auckland, New Zealand: ADIS Press Limited, 1990, p. 17.

2. Pizzorno, J. E., and Murray, M. T., editors. *A Textbook of Natural Medicine.* Seattle, Wash.: Bastyr College Publications, 1991, p. V Ginkgo 7.

## POTENTIAL SIDE EFFECTS

1. Pizzorno, J. E., and Murray, M. T., editors. *A Textbook of Natural Medicine.* Seattle, Wash.: Bastyr College Publications, 1991, p. V Ginkgo 7.

## CONCLUSION

1. DeFeudis, F. V. *Ginkgo Biloba Extract (EGb 761): Pharmacological Activities and Clinical Applications.* Paris: Elsevier, 1991, p. 155.

# INDEX